Emergency Vehicle Visibility and Conspicuity Study

FA-323/August 2009

FEMA

U.S. Fire Administration
Mission Statement

We provide National leadership to foster a solid foundation for local fire and emergency services for prevention, preparedness and response.

PREFACE

The U.S. Fire Administration (USFA) would like to acknowledge the U.S. Department of Justice (DOJ), National Institute of Justice (NIJ), for providing the substantial support necessary to perform this research and to develop this report.

This report was prepared through a Cooperative Agreement between USFA and the International Fire Service Training Association (IFSTA) at Oklahoma State University (OSU). IFSTA and its partner OSU Fire Protection Publications has been a major publisher of fire service training materials since 1934, and through its association with the OSU College of Engineering, Architecture, and Technology, it also conducts a variety of funded technical research on fire service, fire prevention, and life safety issues.

The extensive information provided in this report would not have been possible without the dedication and efforts of the following people assigned to this project:

- Nancy J. Trench–Project Administrator
- Michael A. Wieder–Principal Investigator/Project Manager
- Cindy Finkle–Editor and Proofreader
- Missy Hannan–Senior Graphic Designer
- Adam K. Thiel–FACETS LLP Writer
- Kevin M. Roche–FACETS LLP Principal Researcher

The content of this report was directed and reviewed by a group of subject matter experts and organizational representatives who have extensive knowledge and interest in this topic. The USFA would like to thank the following individuals and organizations for providing this oversight:

- Dick Ashton–International Association of Chiefs of Police (IACP)
- Bill Ballantyne–Fire Apparatus Manufacturers Association (FAMA)
- Dave Bryson–DOT National Highway Traffic Safety Administration (NHTSA), Office of EMS
- Vanessa Castellanos–DOJ/NIJ
- Richard Duffy–International Association of Fire Fighters (IAFF)
- Michael Flannagan–University of Michigan Transportation Research Institute (UMTRI)
- Kenn Fontenot–National Volunteer Fire Council (NVFC)
- Thomas Hughes–IFSTA/Fire Protection Publications
- John McDonald–General Services Administration (GSA)
- Larry McKenna–USFA
- Brian Montgomery–DOJ/NIJ
- Paul Moore–National Institute for Occupational Safety and Health (NIOSH)
- Markus Price–DOT, NHTSA
- Jack Sullivan–International Association of Fire Chiefs (IAFC) and Cumberland Valley Volunteer Fireman's Association (CVVFA)
- Bill Troup–USFA
- Robert Tutterow–National Fire Protection Association (NFPA)
- Keith Williams–DOT, NHTSA, Enforcement and Justice Services Division

A number of organizations and companies also provided valuable assistance during the preparation of this report. This included information on their products and access to their facilities.

- 3M–St. Paul, MN
- Arizona Department of Public Safety–Phoenix, AZ
- Avery–Dennison–Chicago, IL
- City of Boston EMS–Boston, MA
- Reflexite Americas–New Britain, CT
- Rosenbauer America–Wyoming, MN
- University of Michigan Transportation Research Institute–Ann Arbor, MI

TABLE OF CONTENTS

Preface 1

Table of Contents 3

Index of Tables and Figures 4

Executive Summary 5

Key Findings and Opportunities 6

Introduction 7

Methodology 7

      Literature Review 8

      Site Visits 8

      Expert Panel 8

Background 9

      Visibility 10

      Conspicuity 10

      Recognition/Identification 11

      Action 11

Science Overview 11

      Retroreflectivity Explained 11

      Retroreflective Technology 13

International Best Practices 14

U.S. Emergency Vehicle Standards 15

      NFPA 1901, *Standard for Automotive Fire Apparatus* 17

      Federal Specification for the Star-of-Life Ambulance 18

Findings 18

      Key Findings 19

Opportunities 23

      Contour Markings 23

Conclusion 26

Photo Credits 26

References 27

Appendix A. FHWA Reflective Sheeting Identification Guide (FHWA, 2005) 38

Appendix B. Chevrons on the Rear of Fire Apparatus:
  The Background (Tutterow, 2008) 40

Appendix C. NFPA 1901, Section 15.9.3 et seq. (NFPA, 2009) 42

INDEX OF TABLES AND FIGURES

Table 1– Key Findings...6

Table 2–Opportunities..6

Table 3–Site Visits..9

Table 4–Expert Panelists and Represented Organizations..9

Figure 1–Types of reflection...11

Figure 2–Retroreflectivity and distance...12

Figure 3–Effect of entrance and observation angles on reflectivity 13

Figure 4–Two principle methods for creating retroreflective properties 13

Figure 5–Retroreflective Battenburg pattern...15

Figure 6–Retroreflective Half-Battenburg pattern 15

Figure 7–High-conspicuity for U.K. police motorcycles 16

Figure 8–U.K. Ambulance livery..16

Figure 9–U.S adaptation of Battenburg pattern 16

Figure 10–Arizona DPS conspicuity treatment...17

Figure 11–NFPA 1901-compliant fire apparatus chevrons 18

Figure 12–Retroreflective striping on GSA-compliant ambulance..19

Figure 13–Contour markings on large vehicles...25

Figure 14–Edge markings on a patrol car 25

Figure 15–Retroreflective logos and emblems 26

EXECUTIVE SUMMARY

Over the past decade, numerous law enforcement officers, firefighters, and emergency medical services (EMS) workers were injured or killed along roadways throughout the United States. In 2008, as with the prior 10 years, more law enforcement officers died in traffic-related incidents than from any other cause; National Law Enforcement Officers Memorial (NLEOM, 2008) over the past 12 years, an average of one officer per month was struck and killed by a vehicle in the United States. (FBI, 2007) Preliminary firefighter fatality statistics for 2008 reflect 29 of 114 firefighters killed on duty perished in motor vehicle crashes, (USFA, 2009a) similar to figures posted in previous years. According to a 2002 study (Maguire, et al.) that aggregated data from several independent sources, at least 67 EMS providers were killed in ground transportation-related events over the 6 years from 1992 to 1997.

These sobering facts clearly demonstrate the importance of addressing vehicle characteristics and human factors for reducing the morbidity and mortality of public safety personnel operating along the Nation's highways and byways. Studies conducted in the United States and elsewhere suggest that increasing emergency vehicle visibility and conspicuity holds promise for enhancing first responders' safety when exposed to traffic both inside and outside their response vehicles (e.g., patrol cars, motorcycles, fire apparatus, and ambulances).

This report, produced in partnership between the U.S. Fire Administration (USFA) and the International Fire Service Training Association (IFSTA), with support from the U.S. Department of Justice (DOJ), National Institute of Justice (NIJ), analyzes emergency vehicle visibility and conspicuity with an eye toward expanding efforts in these areas to improve vehicle and roadway operations safety for all emergency responders. Emphasis in this report is placed on passive visibility/conspicuity treatments; additional studies are underway on active technologies such as emergency vehicle warning lighting systems. (USFA, 2009b)

A number of key findings were developed from the examination performed for this report. Principal among these findings is the salient need for additional research on emergency vehicle visibility and conspicuity in the United States, with particular emphasis on the interaction between civilian drivers and emergency vehicles during responses and on incident scenes; other key findings are summarized in **Table 1** on the following page.

Despite meaningful limitations, the existing visibility/conspicuity research, combined with passenger vehicle lighting and human factors, evokes several potential opportunities for improving the safety of emergency vehicles in the United States using readily available products. They are summarized in **Table 2**.

Table 1–Key Findings

The increased use of retroreflective materials holds great promise for enhancing the conspicuity of emergency vehicles.

Both visibility and recognition are important facets of emergency vehicle conspicuity.

The use of contrasting colors can assist drivers with locating a hazard amid the visual clutter of the roadway.

Fluorescent colors (especially fluorescent yellow-green and orange) offer higher visibility during daylight hours.

There is limited scientific evidence that drivers are "drawn into" highly-visible emergency vehicles.

It is theoretically possible to "over-do" the use of retroreflective materials and interfere with drivers' ability to recognize other hazards.

Effectiveness of the "Battenburg" pattern in the UK appears primarily related to its association with police vehicles in that country.

Table 2–Opportunities

Outline vehicle boundaries with "contour markings" using retroreflective material, especially on large vehicles.

Concentrate retroreflective material lower on emergency vehicles to optimize interaction with approaching vehicles' headlamps.

Consider (and allow) the use of fluorescent retroreflective materials in applications where a high degree of day-/night-time visibility is desired.

Using high-efficiency retroreflective material can improve conspicuity while reducing the amount of vehicle surface area requiring treatment.

For law enforcement vehicles, retroreflective material can be concentrated on the rear to maintain stealth when facing traffic or patrolling.

Applying distinctive logos or emblems made with retroreflective material can improve emergency vehicle visibility and recognition.

Introduction

The importance of addressing vehicle characteristics and human factors to help positively affect the safety of emergency workers operating along the Nation's roadways is starkly established by first responders' morbidity and mortality experience. Over the past decade, numerous law enforcement officers, firefighters, and EMS workers were injured or killed in roadside crashes throughout the United States.

In 2008, as with the prior 10 years, more law enforcement officers died in traffic-related incidents than from any other cause; (NLEOM, 2008) over the past 12 years, an average of one officer per month was struck and killed by a vehicle in the United States. (FBI, 2007) Preliminary firefighter fatality statistics for 2008 show that 29 of 114 firefighters killed on duty perished in motor vehicle crashes, (USFA, 2009a) similar to figures posted in previous years. According to a 2002 study (Maguire, et al.) that aggregated data from several independent sources, at least 67 EMS providers were killed in ground transportation-related events over the 6 years from 1992 to 1997.

Previous studies conducted across the United States and in other countries suggest that steps to improve emergency vehicle visibility and conspicuity hold promise for enhancing first responders' safety when exposed to traffic both inside and outside their response vehicles (e.g., patrol cars, motorcycles, fire apparatus, and ambulances).

This study explored commercially available vehicle conspicuity products with the goal of increasing their use in helping to enhance emergency vehicle visibility and roadway operations safety for both emergency responders and the general public. Features such as retroreflective striping and chevrons, high-visibility paint, built-in passive lighting, and other reflectors were examined. Emphasis in this report is placed on passive visibility/conspicuity treatments; companion studies are underway on active technologies such as emergency vehicle warning lighting systems. (USFA, 2009b) U.S. and international best practices for emergency vehicle visibility/conspicuity were also assessed to provide information useful for developing standards published by the NIJ, National Fire Protection Association (NFPA), General Services Administration (GSA), and other stakeholder groups.

The report's major finding is the urgent need for additional research specific to emergency vehicle visibility/conspicuity in the United States. A number of other key findings are discussed with implications for deploying existing conspicuity treatments, as well as developing future technologies, standards, and safe operating procedures.

This study identified several possibilities for enhancing emergency vehicle visibility/conspicuity using readily available products. Given the daily risks faced by law enforcement, fire service, and EMS personnel along U.S. roadways, and despite the limitations of the existing literature, these immediate opportunities are detailed in this report.

Methodology

This section describes the multiple methods used for researching this report. An extensive literature review was conducted to gather information from sources across the United States and other countries. Site visits were performed at several manufacturers of retroreflective sheeting products, with emphasis on emergency vehicle installations (versus other applications such as traffic signs or personal protective equipment [PPE]). Additional visits and interviews took place with representative user agencies (law enforcement, fire service, EMS), a leading academic research institute, and a fire apparatus manufacturer. The researchers also convened a panel of subject matter experts on August 11, 2008, in Linthicum, MD; panel members provided individual input on key findings and potential opportunities for improving U.S. emergency vehicle visibility and conspicuity.

Literature Review

The academic and scientific literature reviewed for this study generally falls into three distinct categories 1) passenger and commercial vehicle visibility/conspicuity, 2) the efficacy of retroreflective sheeting material used for traffic signs and other stationary applications, and 3) emergency vehicle visibility/conspicuity.

The majority of the relevant literature on vehicle visibility and conspicuity revolves around the development, adoption, implementation, and evaluation of U.S. (Green et al., 1979; Burger & Smith, 1987; Olson et al., 1992; Morgan, 2001; Sullivan & Flannagan, 2004; Sullivan, 2005; Donelson & Ayers, 2007) and European (Cook et al., 1999; Schmidt-Clausen, 2000; TUV Rheinland Group, 2004; Richardson & Lawton, 2005) regulatory requirements/practices for marking commercial truck trailers and heavy goods vehicles (HGVs), as they are called in Europe with retroreflective material. Since heavy fire apparatus (engines/pumpers, ladder trucks, heavy rescue squads, etc.) possess physical characteristics similar to large commercial trucks, this research is thought applicable for some emergency vehicles.

The contemporary literature on retroreflective sheeting materials used for traffic signs and other stationary installations is robust. (Anders, 2000; Hawkins et al., 2000; Chrysler et al., 2002; Chrysler et al., 2003; Carlson & Urbanik, 2004; Gates & Hawkins, 2004; Rogoff et al., 2005; Sivak et al., 2006; Amjadi, 2008) The same types of retroreflective products are used in emergency vehicle marking applications and are therefore relevant to this report.

While the literature identifies emergency vehicle visibility/conspicuity as a major concern in the United States, (CVVFA, 1999; AZ DPS, 2003; IACP, 2004; USFA, 2004; Burbank, 2007; McCann et al., 2008; NFPA, 2008; Ridenour et al., 2008; NFPA, 2009; USFA, 2009b), true empirical research specific to U.S. emergency vehicle visibility and conspicuity is almost nonexistent. (Interview with Dr. Michael Flannagan, 2008) This represents a meaningful gap, since recommendations and best practices stemming from research performed in countries like the United Kingdom (Thomas, 1998; Harrison, 2004; Harrison, 2006; BSI, 2007) may not be generalizable to the United States, particularly as they relate to different traffic safety cultures (Sivak et al., 1989; AAA, 2007; McNeely & Gifford, 2007; Williams & Haworth, 2007) and civilian drivers' interaction with emergency vehicles. In fact, there is very little research at all on how drivers perform when faced with responding or parked emergency vehicles in the U.S. and elsewhere. (Tijerina et al., 2003; interview with Dr. Michael Flannagan, 2008)

Site Visits

The goal of the site visits displayed in **Table 3** was threefold 1) gather information on the nature of the emergency vehicle visibility/conspicuity problem, 2) examine commercially available products with the potential for improving the conspicuity of emergency vehicles, and 3) identify best practices for enhancing emergency vehicle visibility and conspicuity.

Expert Panel

The expert panel, convened in Linthicum, MD, on August 11, 2008, provided substantial input to this report. Participants reviewed the nature of the problem from a multidisciplinary perspective (law enforcement, fire, EMS, regulatory, academic/scientific); identified additional information sources and research; discussed initial findings; and talked about preliminary opportunities for improving emergency vehicle visibility and conspicuity.

As reflected in **Table 4**, subject matter experts for the panel were drawn from many components of the emergency vehicle visibility/conspicuity stakeholder community across the United States.

Table 3–Site Visits	
3M	St. Paul, Minnesota
Arizona Department of Public Safety	Phoenix, Arizona
Avery-Dennison	Chicago, Illinois
Boston EMS	Boston, Massachusetts
Reflexite Americas	New Britain, Connecticut
Rosenbauer America	Wyoming, Minnesota
University of Michigan Transportation Research Institute (UMTRI)	Ann Arbor, Michigan

Table 4–Expert Panelists and Represented Organizations	
Bill Ballantyne	Fire Apparatus Manufacturers Association
John McDonald	General Services Administration
Dick Ashton	International Association of Chiefs of Police
Jack Sullivan	International Association of Fire Chiefs
Thomas Hughes	International Fire Service Training Association/Fire Protection Publications
Robert Tutterow	National Fire Protection Association
Vanessa Castellanos	National Institute of Justice
Brian Montgomery	National Institute of Justice
Paul Moore	National Institute for Occupational Safety and Health
Markus Price	National Highway Traffic Safety Administration
Keith Williams	National Highway Traffic Safety Administration, Enforcement and Justice Services Division
Dave Bryson	National Highway Traffic Safety Administration, Office of EMS
Kenn Fontenot	National Volunteer Fire Council
Michael Flannagan	University of Michigan Transportation Research Institute
Larry McKenna	United States Fire Administration
Bill Troup	United States Fire Administration
Kevin Roche	FACETS Consulting, LLP
Adam Thiel	FACETS Consulting, LLP

Background

Visibility and conspicuity are only two aspects of ensuring emergency vehicle safety while in traffic and parked along roadways. Recognition and driver action(s) are also important facets of an extremely complicated, interdependent, and largely ill-defined system that includes multiple vehicles, drivers, their culture(s), and the environment. This section provides background essential for understanding this complex system.

Visibility

A number of interrelated factors affect the visibility of an emergency vehicle to adjacent drivers both during a response and while parked at an incident scene. These variables include the vehicle's size, color scheme (also called a "livery"), passive conspicuity features such as marker lamps and retroreflective striping, and the presence/operation of active warning devices including emergency lighting systems or audible sirens and horns. Environmental conditions also influence visibility; chief among these are time-of-day, ambient lighting, weather, and the presence of driver distractions or visual clutter in the surroundings.

While a high degree of visibility is usually a desirable characteristic for emergency vehicles, there are times when public safety personnel do not want their vehicles to be readily visible. Law enforcement officers, for example, may actually want to be almost invisible to other drivers when conducting certain enforcement or patrol activities. There is also a point-of-view (discussed later in this report) that, while parked off the roadway at an incident, emergency vehicles should reduce their conspicuity to avoid being hit by drivers who are potentially attracted to activated warning devices.

The dichotomy between the need for emergency vehicles to be highly visible under some scenarios, and less visible under others, creates different conspicuity requirements for emergency vehicles than for vehicles like school buses where high visibility is always desired.

Conspicuity

Conspicuity refers to, "the ability of a vehicle to draw attention to its presence, even when other road users are not actively looking for it." (Cook et al., 1999, p. 3) All vehicles sold in the United States come factory-equipped with certain legally-required devices, making them conspicuous to a basic degree. This includes headlamps, parking lamps, brake (stop) lamps, tail lamps, marker lamps, signal lamps, reflex reflectors, etc. Beyond any original equipment manufacturer-(OEM)-installed conspicuity features, emergency vehicle visibility is meaningfully affected to the extent that a vehicle is deliberately made conspicuous, or inconspicuous, to other drivers. Unmarked, undercover, and low-profile vehicles are at one end of the conspicuity spectrum; while a brightly colored, perhaps even fluorescent, fire apparatus equipped with multiple active warning systems and passive conspicuity treatments anchors the other end of the continuum.

Historically, emergency vehicle operators primarily relied on active signaling, using various mechanical devices, to enhance their vehicles' visibility and conspicuity while responding to, and on the scene of, emergency incidents. These technologies include emergency warning lights and audible systems designed to attract surrounding drivers' attention. While active devices will likely always be needed to promote emergency vehicle visibility/conspicuity, passive treatments using retroreflective sheeting and other materials are increasingly being used to complement lights and sirens. (Warning lighting systems are the subject of another USFA report, **Study of Emergency Vehicle Warning Lighting Systems**, prepared under its **Emergency Vehicle Safety Initiative**.) (USFA, 2009b)

Studies conducted in the United States and other countries suggest efforts to increase emergency vehicle conspicuity using passive treatments hold potential for enhancing emergency responders' safety when exposed to traffic during responses or at the roadside. However, given the number of variables present in a wide range of driver-emergency vehicle interaction scenarios, it is vital to recognize the optimal combination of conspicuity markings and active warning systems for every possible situation probably does not exist. In fact, the best choices for conspicuously marking stopped emergency vehicles could be quite different from the best choices for enhancing the visibility of those same vehicles while in motion. (Tijerina et al., 2003; Donelson & Ayers, 2007)

Recognition/Identification

The purpose of making any vehicle conspicuous goes further than simply enhancing its visibility. While catching the eye of another driver is the "first thing," the larger goal is to help provide other drivers with information about a vehicle's presence, size, position, speed, and direction of travel. As a driver, the critical objectives of conspicuity are to 1) clearly broadcast your own aims, and 2) easily recognize surrounding drivers' intentions, enabling the appropriate action to avoid a collision. For emergency vehicles, recognition and identification are likely very important aspects of promoting this so-called "cognitive conspicuity." (Tijerina et al., 2003)

Action

Beyond recognizing the presence of an emergency vehicle, civilian drivers must know what action(s) to take after identifying it. There are two opposing viewpoints on this front 1) it might be preferable for drivers to simply recognize the presence of something they need to avoid, without any specific information on why they should do so (avoiding distraction as drivers turn their attention away from driving), or 2) drivers should be able to quickly recognize and specifically identify basic types of emergency vehicles, since doing so will help them determine the right course of action to avoid impeding a response or safely negotiate an emergency scene.

The previously identified lack of research on U.S. drivers' interaction with emergency vehicles makes it extremely difficult to answer fundamental questions about the ultimate effectiveness of visibility/conspicuity treatments on the safety of emergency vehicles in the overall traffic system.

Science Overview

This section provides a brief overview of the science behind reflection and retroreflective technology. Generally speaking, every surface reflects light to some degree; this physical property allows us to see objects when they do not emit their own light source. There are two principal types of reflection 1) **diffuse** reflection happens when light strikes a rough surface and is reflected or "diffused" in all directions, and 2) **specular**, or "mirror-like," reflection occurs when light strikes a smooth surface (like mirrored glass) and the reflected light is returned along the same angle as the incoming path, returning an image along with the reflected light. A third type of reflection, **retroreflection**, occurs when a surface is specially engineered to reflect light back to its origin. **Figure 1,** depicts these three types of reflection using a simple graphic from the Federal Highway Administration (FHWA).

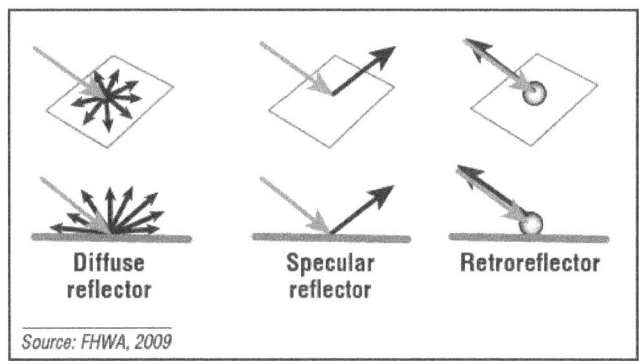

Source: FHWA, 2009

Figure 1—Types of reflection.

Retroreflectivity Explained

Within the law enforcement, fire service, and EMS disciplines, emergency warning lights are the predominant method for making emergency vehicles conspicuous. The USFA (2009b) and other organizations continue producing research to refine emergency vehicle lighting systems and enhance their effectiveness. Another method of advancing night-time emergency vehicle conspicuity is through the increased use of retroreflective materials. Retroreflective materials are defined as those that (re)direct incoming light back to the viewer, such as the driver of a vehicle approaching a roadside incident scene.

Retroreflectivity doesn't just happen; for materials to exhibit their retroreflective properties, an external light source is needed. While emergency vehicles carry their own light sources in the form of headlamps, marker lamps, and emergency warning lights, they also depend on light from other vehicles' headlamps for visibility. The degree to which a retroreflective object (including an emergency vehicle treated with retroreflective striping) reflects light back to its origin depends on the amount of incoming light hitting the retroreflective surface(s), as displayed in **Figure 2**.

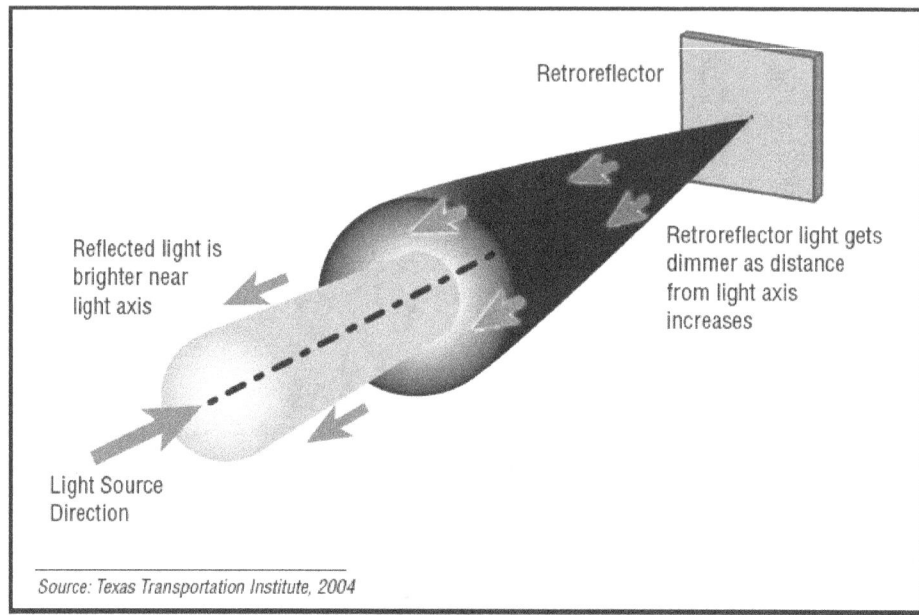

Figure 2—Retroreflectivity and distance.

Reflected light is brighter near light axis

Retroreflector

Retroreflector light gets dimmer as distance from light axis increases

Light Source Direction

Source: Texas Transportation Institute, 2004

The intensity of light emitted from vehicle headlamps and other sources is measured in candelas (cd). The amount of light striking a surface is expressed as illuminance and measured in lux (lux). Light reflected back to an observer, seen as "brightness" and called luminance, is measured in candelas per meter-squared (cd/m^2). The coefficient of retroreflection (RA) is the ratio of light reflected (luminance) from a retroreflective surface to the illuminance, as described by the formula:

$$R_A = cd/m^2$$

The R_A is a relative measure of efficiency for a given retroreflector at a specific viewing geometry.

The viewing geometry of a retroreflector is a function of two angles 1) the angle that incoming light strikes the target (such as a traffic sign, vehicle, person, or other object), the entrance angle (ß), and 2) the angle where light reflected back from the target is observed, the observation angle (a). Changing the angles at which a retroreflective target accepts incoming light and reflects it back to the viewer(s) changes the visibility/conspicuity of retroreflective materials, as displayed in **Figure 3**.

This phenomenon has crucial implications for the use of retroreflective materials on emergency vehicles since, unlike a fixed traffic sign where the expected viewing geometry is largely predictable, in a vehicle-mounted application the relative positions of target and observer are continually changing, thus changing the viewing geometry.

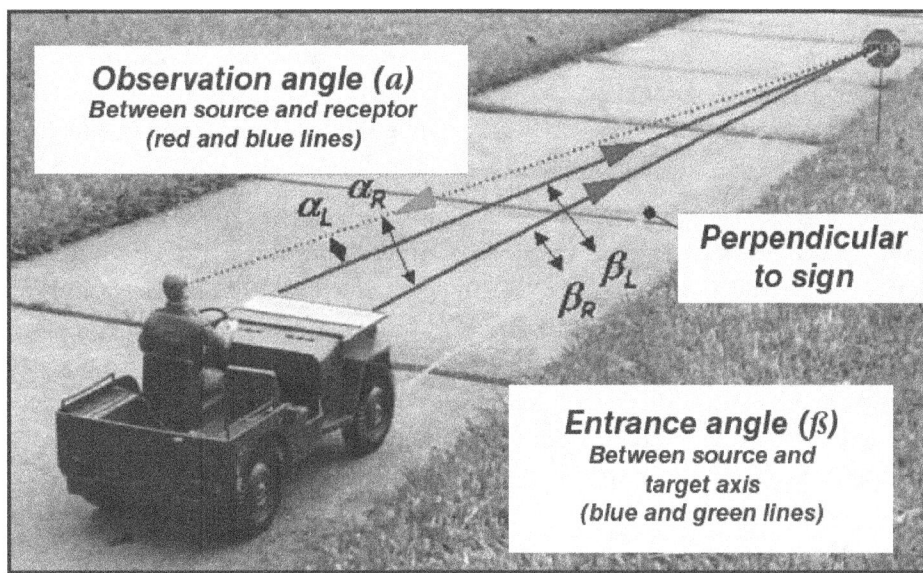

Observation angle (*a*)
Between source and receptor
(red and blue lines)

α_L α_R

β_L
β_R

Perpendicular
to sign

Entrance angle (*ß*)
Between source and
target axis
(blue and green lines)

Figure 3—Effect of entrance and observation angles on reflectivity.

Source: FHWA, 2009

Retroreflective Technology

Manufacturers of retroreflective sheeting products are constantly developing new technology, and refining existing materials, to increase retroreflective efficiency while providing other characteristics (e.g., low cost, ease-of-installation, flexibility, durability, color selection, customization, etc.) demanded by customers.

Contemporary retroreflective sheeting products are made by applying microscopic glass spheres (beads), or engineered microprisms (cubes), to a more-or-less flexible substrate in an arrangement promoting internal reflection and the efficient return of incoming light back to its source. The surfaces of the spheres or microprisms are further engineered to promote retroreflectivity through polishing and/or coating with metallic or nonmetallic materials to give a "mirror" effect. The optical properties and arrangement of these two geometric structures combine to direct incoming light back toward its origin, as displayed in **Figure 4**.

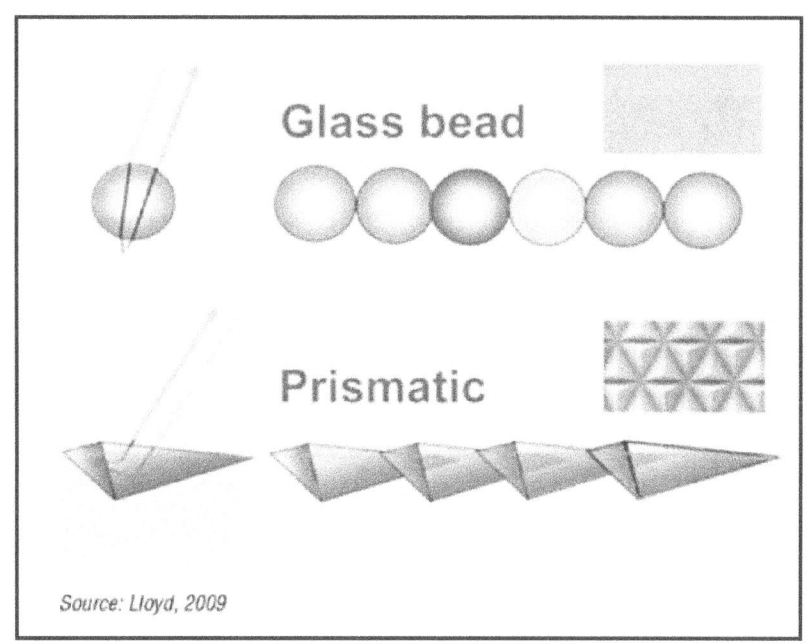

Glass bead

Prismatic

Source: Lloyd, 2009

Figure 4— Two principle methods for creating retroreflective properties.

The amount of light returned to a source by a given retroreflective material, at a specific viewing geometry, is known as its retroreflective efficiency.

Retroreflective efficiency generally ranges from 7 to 14 percent for sheeting made with glass beads, up to 32 percent for "truncated-cube" microprismatic sheeting, and 58 percent for "full-cube" microprismatic sheeting. (minimumreflectivity.org, 2009) It is important to note that a higher retroreflective efficiency does not necessarily imply a "better" material. Different sheeting types have other properties that make them more or less suitable for a particular application (e.g., cost, flexibility, durability, ease-of-installation, color, customization, etc.). Furthermore, while beaded sheeting is relatively less efficient than microprismatic sheeting, the use of spheres versus cubes helps maintain its performance over a wider range of viewing geometries.

Retroreflective sheeting materials generally are designed and tested for use on traffic control devices such as signs, barriers, and cones, although the same products are used in other applications (i.e., on clothing or vehicles). The American Society for Testing and Materials (ASTM) maintains the **ASTM D4956 Standard Specification for Retroreflective Sheeting for Traffic Control**. (ASTM International, 2007) The current (2007) edition of ASTM D4956 specifies 10 different types (I, II, III...X) of retroreflective sheeting materials. A complete list is included as Appendix A to this report. A higher ASTM type-number does not necessarily reflect a greater retroreflective efficiency or a better product. The type designations simply identify materials that meet different performance specifications.

International Best Practices

During the past 10 years, the United Kingdom government researched and deployed a set of visibility/conspicuity standards now used on law enforcement vehicles throughout the country. Efforts to develop conspicuity specifications in the United Kingdom were undertaken with several objectives in mind:

- recognizable at a distance from 200 to 500+ meters (650 to 1,650+ feet);
- assist with high-visibility policing;
- readily identifiable nationally as a police vehicle, with room for local markings; and
- acceptable to at least 75 percent of the staff using it (Harrison, 2006).

Anecdotal evidence suggests these standards are being emulated, to various degrees, by other public safety services (fire, EMS, etc.) across the United Kingdom, in other Nations (e.g., Australia, South Africa, Sweden, and New Zealand), and some places in the United States. This section describes notable features of the liveries used in the United Kingdom, as well as some considerations in their development.

The **Specification for the Livery on Police Patrol Cars** (Thomas, 1998) was designed to make police vehicles operating along high-speed roadways in the United Kingdom "visible throughout the day and night and be clearly identifiable as a police car." (p. 1) This design considers a minimum viewing distance of 500 meters (1,650 feet) under weather conditions including "rain, mist, etc.," with night-time illumination provided by an approaching vehicle with normal headlights. (Thomas, 1998, p. 1) In addition to retroreflective chevrons on the rear of the patrol car, this livery also requires a retroreflective "Battenburg" (also seen as "Battenberg" or "harlequin") pattern along the sides, ostensibly to improve both day- and night-time conspicuity and recognition as a police vehicle, as seen in **Figure 5**.

In 2004, the United Kingdom Home Office Scientific Development Branch published a subsequent specification detailing a "high-conspicuity" livery for police vehicles used in cities and towns. (Harrison, 2004) In addition to the "full-Battenburg" scheme used on patrol cars primarily assigned to high-speed roadways, the 2004 document specifies a "half-Battenburg" pattern for patrol vehicles deployed in the urban environment. An example of this scheme is displayed in **Figure 6**.

Illustrated in **Figure 7** on page 16, the United Kingdom high-conspicuity livery for police motorcycles (Harrison, 2006) demonstrates the consideration placed on the motor officer's visibility/conspicuity as part of the overall scheme for making the vehicle (and rider) highly visible to surrounding traffic.

Figure 5—Retroreflective Battenburg pattern.

Figure 6—Retroreflective Half-Battenburg pattern.

The recent edition of the British Standards Institute (2007) (**BS EN 1789:2007 Standard for Medical Vehicles and their Equipment Road Ambulance**) requires the application of "micro-prismatic reflective material" (p. 41) for visibility at night. While this standard does not specify a Batternburg pattern, many EMS agencies in the United Kingdom have adapted the full- or half-Battenburg scheme to ambulances, illustrated in **Figure 8** on page 16, using different colors (typically yellow and green) than their law enforcement counterparts.

Some public safety agencies in the United States have extrapolated the United Kingdom high-conspicuity liveries, including the Battenburg pattern, to their vehicles; examples are displayed in **Figure 9** on page 16.

U.S. Emergency Vehicle Standards

By contrast to the United Kingdom, there is currently no evidence of a U.S. industry standard for the visibility/conspicuity of law enforcement vehicles. As a best practice, however, many U.S. law enforcement agencies apply retroreflective treatments to patrol cars, motorcycles, and other vehicles. The Arizona Department of Public Safety (AZ DPS) is one of several leaders in this area. An example of the retroreflective conspicuity treatment applied to their highway patrol vehicles is provided in **Figure 10**.

Recent editions of national standards for U.S. fire apparatus and ambulances require the increased (from previous versions) requirements for the application of retroreflective striping and markings to enhance visibility and conspicuity. (NFPA, 2009; GSA, 2007) The balance of this section describes features of these standards and some issues surrounding their adoption.

Figure 7—High-conspicuity for U.K. motorcycles.

Figure 8—U.K. Ambulance livery.

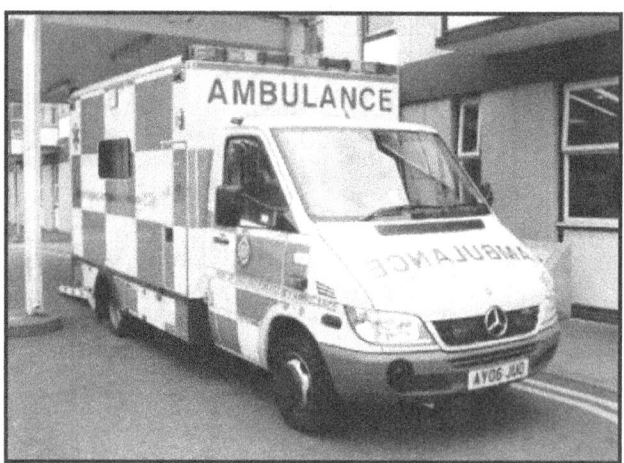

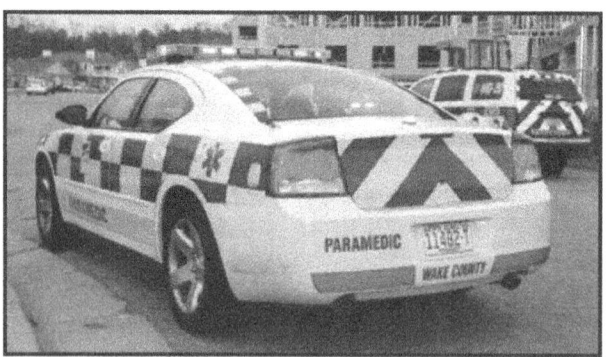

Figure 9—U.S. adaptation of Battenburg pattern.

NFPA 1901, Standard for Automotive Fire Apparatus

The **NFPA 1901, Standard for Automotive Fire Apparatus, 2009 edition,** (NFPA, 2009) is a voluntary national standard developed through a consensus-based process involving multiple stakeholders such as manufacturers, users, installers/ maintainers, labor, applied research/testing laboratories, enforcing authorities, insurers, consumers, and special experts in relevant fields. While compliance with NFPA standards is voluntary, unless adopted as a code/ordinance/regulation by the authority having jurisdiction (AHJ), manufacturers typically comply with the latest version of the relevant standard(s) to limit legal liability and ensure product marketability.

Effective January 1, 2009, section 15.9.3 et seq. of the **NFPA 1901, Standard for Automotive Fire Apparatus, 2009 edition,** (NFPA, 2009) requires retroreflective striping in multiple locations, including at least:

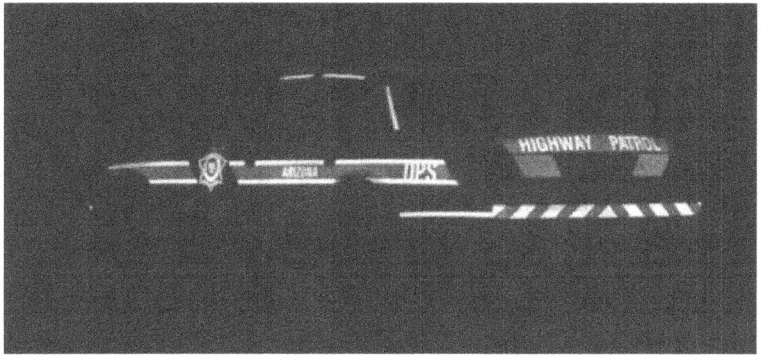

Figure 10— Arizona DPS conspicuity treatment.

- fifty percent of the cab and body length on each side (excluding pump panels) with 4-inch wide striping;

- twenty-five percent of the front width of the apparatus with 4-inch wide striping;

- fifty percent of rear-facing vertical surfaces (excluding pump panels not covered by a door) in a 45-degree down-and-away "chevron" pattern with alternating red and yellow/fluorescent yellow/fluorescent yellow-green 6-inch stripes. (NFPA, 2009, p. 47)

While the application of retroreflective striping and trim is nothing new for fire apparatus, the amount demanded by the 2009 edition of NFPA 1901 drew a great deal of comment from the United States fire service, particularly related to the chevrons used for marking rear-facing surfaces. The most controversial issue was the requirement for a standardized red and yellow/fluorescent yellow/fluorescent yellow-green pattern for the chevrons, as illustrated in **Figure 11** on page 18. Fire departments across the United States use myriad color schemes based largely on the traditions and creativity of their members and sometimes passed down through generations. Many commenters on the proposed 2009 edition of NFPA 1901 supported the notion of retroreflective chevrons, but felt strongly about selecting their own striping colors to match the rest of the vehicle's livery. (Personal conversation with Mr. Robert Tutterow, 2008)

The NFPA Technical Committee on Fire Department Apparatus ultimately decided that standardization across the United States was the best course of action; their rationale is explained in the Winter 2008 edition of the NFPA Fire Service Section Newsletter (Tutterow, 2008), reproduced in its entirety as Appendix B of this report.

Figure 11—NFPA 1901-compliant fire apparatus chevrons.

Beyond its design stipulations, **NFPA 1901 cites section 6.1.1 of ASTM D4956 Standard Specification for Retroreflective Sheeting for Traffic Control** (ASTM International, 2007) in detailing the performance/colors of retroreflective materials used for the required conspicuity treatment. During the research for this report, an issue with the colors identified in the NFPA 1901 standard was noted. Specifically, while NFPA 1901 allows the use of "fluorescent yellow-green," for the chevron pattern, strict adherence to the cited table ("color box") in the ASTM D4956 standard would disallow it. (Personal conversation with Mr. Tom Bliss et al., 2008) The need for synchronization was discussed with members of the NFPA Technical Committee on Fire Department Apparatus during the expert panel meeting.

Federal Specification for the Star-of-Life Ambulance

On August 1, 2007, the U.S. General Services Administration (GSA) published the latest edition of its **KKK-A-1822F Federal Specification for the Star-of-Life Ambulance**. (GSA, 2007) While it is not completely applicable to every ambulance operated in the United States, the GSA standard (also known as the "Triple-K spec") outlined in this section of the report sets forth the minimum requirements followed by OEMs nationwide.

Conspicuity is integral to the basic ambulance livery described in section 3.16 et seq. of the 2007 edition of the GSA specification:

> The exterior color of the ambulance shall be gloss white in combination with a solid uninterrupted orange stripe and blue lettering and emblems...The orange stripe shall not be less than 6 inches wide, nor more than 14 inches wide and shall encircle the entire ambulance body at the belt line below the bottom edge of cab windows but may exclude the front of the hood panel. The orange stripe shall be reflective tape. (GSA, 2007, p.51)

Beyond the basic color scheme, KKK-A-1822F specifies retroreflective sheeting for all emblems (star-of-life) and "ambulance" markings:

> The material for the emblems and markings shall be applied using reflective material that has a coefficient of retroreflection measured in accordance with ASTM E 810 of 100 for White and 10 for Blue using 4° entrance angle and a 0.2° observation angle. (GSA, 2007, p.52)

An ambulance meeting the basic tenets of the Triple-K specification is displayed in **Figure 12**.

Findings

The present understanding of emergency vehicle visibility and conspicuity in the United States is extremely limited. Researchers have little comprehension of how civilian drivers actually negotiate emergency scenes and interact with moving or parked fire apparatus, ambulances, and patrol cars; for this reason, there is minimal scientific evidence to directly support any specific recommendations for enhancing U.S. emer-

Figure 12—Retroreflective striping on GSA-compliant ambulance.

gency vehicle conspicuity using retroreflective materials or other passive treatments. Most of the current advances in this area are based on anecdote and extrapolation from studies in other venues that may or may not translate into meaningful safety improvements. In fact, some of these "remedies" might actually prove detrimental. Additional research specific to U.S. emergency vehicle visibility/conspicuity is sorely needed, particularly given the complexity, diversity, and interdependencies of the American traffic system and its driving culture(s).

Key Findings

Notwithstanding the limitations described above, this section presents a number of key findings from the research performed for this report.

Retroreflective Materials

It seems clear that properly applied/maintained retroreflective sheeting materials can effectively increase the night-time visibility and conspicuity of treated objects, as frequently used across the United States in a wide range of traffic control applications. (FHWA, 2007) While generalizing practices (without rigorous evaluation) used in other disciplines and/or countries remains a concern, the current research suggests that leveraging the properties of readily available retroreflective sheeting products, by incorporating them into U.S. emergency vehicle designs, appears promising for enhancing emergency vehicle visibility and conspicuity, especially during dark lighting conditions. (Retroreflectivity is of limited benefit with daylight illumination.)

Several U.S. researchers (Green et al., 1979; Sivak, 1979) evaluated the need to improve night-time vehicle conspicuity and the potential contribution of retroreflectorization toward that goal; their results directed interest at using retroreflective materials for improving the conspicuity of trucks (generally tractor-trailers) and other highway vehicles. Later experiments with large truck retroreflectorization (Burger & Smith, 1987) identified a 15 percent reduction in accidents for trucks with retroreflectors, versus those without. These efforts, and others testing various performance specifications for conspicuity treatments, (Olson et al., 1992) catalyzed the 1992 addition of retroreflective conspicuity standards for U.S. trucks in the **Federal Motor Vehicle Safety Standards (FMVSS), Chapter 108**. (US CFR, 2004) Administered by the National

Highway Traffic Safety Administration (NHTSA), and applicable to all U.S. motor vehicles, the stated purpose of FMVSS 108 is to:

> ...reduce traffic accidents and deaths and injuries resulting from traffic accidents, by providing adequate illumination of the roadway, and by enhancing the conspicuity of motor vehicles on the public roads so that their presence is perceived and their signals understood, both in daylight and in darkness or other conditions of reduced visibility. (US CFR, 2004)

Research performed by NHTSA suggests retroreflective conspicuity treatments applied to U.S. heavy truck trailers since 1992, with a retrofit requirement in 1999, (US CFR, 2004) have been "quite effective" at reducing side-/rear-impact crashes at night. (Morgan, 2001, p.vi) These findings comport with several other studies on retroreflectorized vehicles in European countries. (Cook et al., 1999; Schmidt-Clausen, 2000; TUV Rheinland Group, 2004; Richardson & Lawton, 2005) Citing large-truck crash data collected since the advent of U.S. conspicuity regulations, some researchers have suggested retroreflective conspicuity treatments are of limited value on moving trucks, offering the most benefit when stopped. (Donelson & Ayers, 2007) Additional retrospective evaluations, while unable to reliably identify the specific benefit of the U.S. regulatory scheme in terms of crash reduction for large trucks, point to a significant reduction in vehicle crashes for both light vehicles and trucks in the dark. (Sullivan & Flannagan, 2004; Sullivan, 2005)

Visibility and Recognition

A wide range of factors affect the visibility and recognition of emergency vehicles, including the presence/operation of active warning devices such as lights and sirens; retroreflective conspicuity treatments (at night); lettering and graphics; and color scheme(s).

The lettering used to mark emergency vehicles almost certainly affects the ability of surrounding drivers to recognize them. Multiple studies have demonstrated that retroreflective sheeting type, font style/size, word count, and color are meaningful factors in determining the legibility of traffic signs and vehicle markings; (Aoki et al., 1989; Krull & Hummer, 2000; Schmidt-Clausen, 2000; Chrysler et al., 2002; Gates & Hawkins, 2004; Amjadi, 2008) it seems likely these results extend to emergency vehicles. One U.S. study suggested yellow, white, green, and orange as good choices for promoting the legibility of retroreflective lettering; (Chrysler et al., 2002) while a European study identified red, white, yellow, and green. (Schmidt-Clausen, 2000) However, as Texas Transportation Institute researchers concluded, "The results of this project demonstrate that it is not practical to identify one combination of font, sheeting, and color that optimizes sign performance in all conditions." (Chrysler et al., 2002, p. 33) Beyond lettering, which requires approaching drivers to read it, European studies suggest the use of retroreflectorized logos and graphics (Schmidt-Clausen, 2000) can positively impact the visibility/conspicuity, and likely the recognition, of large vehicles.

Visibility and recognizability are likely influenced by the color scheme(s) in which emergency vehicles are painted/decorated. The literature reviewed for this report identified multiple colors, and combinations/patterns, as beneficial for improving vehicle conspicuity. From these inconclusive results, it seems clear that no single particular color represents the optimal choice for enhancing emergency vehicle visibility/conspicuity under every possible scenario. Beyond the physics and psychology of how different colors are viewed/seen under varying conditions, it seems probable that cultural factors are also salient in terms of how civilian drivers interpret the use of color in traffic safety applications. One obvious example is the association of red and white with a "stop" message; perhaps due to this connection, and the use of red tail lamps on vehicles, several studies identified the color red as a good choice for marking the rear of a vehicle. (Burger et al., 1985; Olson et al., 1992; Cook et al., 1999) Different colors also behave differently in terms of their luminance when applied to objects using different types of retroreflective sheeting. One study judged red, green, blue and orange as "brighter," according to test subjects, than white or yellow. (Aoki et al., 1989)

There is some literature on the selection of different paint colors for emergency vehicles. Solomon (1990) suggested the predominant color scheme on United States fire apparatus change from red to "lime-yellow." (Solomon & King, 1995) The rationale for this proposal is based on the assertion that yellow-green is an easy color for the human eye to discern in both day/night lighting conditions, as well as providing contrast with typical backgrounds. (Solomon & King, 1997) United Kingdom researchers Langham and Rillie (2002) proposed a single-color paint scheme, using fluorescent orange, as the appropriate choice for emergency vehicle visibility/conspicuity under most environmental conditions. Whatever the specific color, research performed for this report suggests what is more important is the ability for drivers to recognize the vehicle for what it is. (Schmidt-Clausen, 2000) The use of a standardized color or paint scheme for certain types of vehicles may be helpful in this regard. (Olson et al., 1992; Thomas, 1998; Harrison, 2004; Harrison, 2006; BSI, 2007) An example is the ubiquitous "yellow school bus" prevalent throughout the United States. These vehicles are instantly recognizable and likely promote immediate behavioral responses by surrounding drivers. Similarly, U.S. Postal Service (USPS) or other mail/delivery trucks painted in a standard color may also prompt drivers to behave in certain ways (i.e., expecting multiple stops at any time). Following this principle, it is a common belief that people are more likely to identify red with a fire apparatus than other colors, regardless of the conditions.

Different marking patterns can also change driver responses. The association of the down-and-away chevron pattern with a "danger" or "slow down" message probably has something to do with its widespread use on traffic barriers, as specified in the U.S. **Manual on Uniform Traffic Control Devices**; (FHWA, 2007) a similar scheme is used to mark immovable roadside objects in many European Nations. Whether or not the chevron pattern actually confuses drivers when it is seen on the back of a moving emergency vehicle is an open question. Some researchers have suggested the best use of chevrons might be deploying them on the rear of emergency vehicles only when they are stopped. (Tijerina et al., 2003) Again, there is a critical need for additional research to identify the best color/pattern selections for emergency vehicles in the United States and elsewhere.

Beyond recognizing the presence of an emergency vehicle, civilian drivers must know what action(s) to take after identifying it. There are two opposing viewpoints on this front 1) it might be preferable for drivers to simply recognize the presence of something they need to avoid, without exact information on why they should do so (avoiding distraction as drivers turn their attention away from driving), or 2) drivers should be able to quickly recognize and specifically identify basic types of emergency vehicles, since doing so will help them determine the right course of action to avoid impeding a response or safely negotiate an emergency scene. Absent targeted research on drivers' interaction with U.S. emergency vehicles, it seems likely the "correctness" of either perspective depends on the situation. (i.e., whether involved emergency vehicles are moving or stopped; the complexity of an incident scene; ambient lighting; weather; road conditions; presence/absence of distractions; degree of visual clutter; etc.).

It is apparent, however, that U.S. emergency services (law enforcement, EMS, and fire) and traffic safety agencies need to better define and educate civilian drivers on the preferred action(s) to take after seeing/ recognizing different types of emergency vehicles. One example of such an education program was created in 2007 by "Move Over, America," a partnership between the National Safety Commission, the National Sheriffs' Association, and the National Association of Police Organizations with full support from the American Association of State Troopers. (www.moveroveramerica.org) To date, all but seven U.S. States have enacted "Move Over Laws," but they vary widely in terms of specific provisions and coverage. (Personal communication with Mr. Dick Ashton, 2009) This and other initiatives can be implemented as part of broader efforts to improve the overall traffic safety culture in the United States (Dula & Geller, 2007).

Contrast

The use of contrasting colors can positively affect conspicuity by assisting drivers with locating a hazard amid the visual clutter of the roadway. There are basically two types of contrast: 1) luminance contrast—the degree to which an object is brighter than its background, and 2) color contrast—the difference in an object's color(s) and those found in its background. (Cook et al., 1999) Contrast is enhanced by using colors not normally found in the environment, including fluorescents.

Fluorescent Colors

The effectiveness of fluorescent colors for enhancing daytime visibility/conspicuity in traffic safety applications is well-established in the literature. (Smith, 1981; Zwahlen & Vel, 1994; Cook et al., 1999; Anders, 2000; Hawkins et al., 2000; Krull & Hummer, 2000; Schieber et al., 2003; Buonarosa & Sayer, 2007) Since fluorescence relies on ultraviolet radiation, fluorescent colors offer no additional benefit at night:

> *Fluorescent colors are brighter than ordinary colors because they are capable of converting light energy that is normally absorbed and wasted to visible light, which in turn reinforces the color in intensity. Hence, there is greater visibility in daylight conditions. (Smith, 1981)*

Extending the concept to emergency vehicles, police patrol car and motorcycle liveries in the United Kingdom liberally employ fluorescent colors to enhance daylight conspicuity. (Thomas, 1998; Harrison, 2004; Harrison, 2006)

The specific color choice may or may not be important with respect to fluorescents, perhaps depending on background characteristics. In a 1994 study, "...fluorescent yellow was found to be best detected and fluorescent orange was found to be best recognized against any of the three backgrounds investigated." (Zwahlen & Vel, abstract) A recent study of traffic safety garments showed no statistical difference in the daytime conspicuity of fluorescent red-orange and fluorescent yellow-green, although fluorescent yellow-green had a significantly higher luminance value, compared to the background, than the fluorescent red-orange. (Buonarosa & Sayer, 2007) Research performed at the Texas Transportation Institute also demonstrated the benefits of fluorescent colors, in this case fluorescent-orange work zone signs, citing greater recognition distance and accurate color perception during the day. (Hawkins et al., 2000)

The "Moth Effect"

There is limited scientific evidence to support the notion that drivers steer toward bright lights, such as those used to increase the visibility of emergency vehicles, as "moths to a flame" (often called the "moth effect" and technically, "phototaxis"). (Interview with Dr. Michael Flannagan, 2008; Green, 2009) Several recent studies, however, suggest that while bright lights may not be the cause, drivers' fixation on roadside objects can cause their steering to drift in the direction of their gaze. (Readinger et al., 2002; Chatziastros et al., 2006) This effect may be more pronounced with other impairments. The implications of these findings on emergency vehicle visibility/conspicuity are unknown, but certainly support the need for additional research on how to design passive conspicuity treatments so they draw drivers' attention enough to induce the appropriate ("stay away") response, without causing the potentially negative results of visual fixation.

Overdoing It

It is *theoretically* possible to "overdo" the use of retroreflective materials and interfere with drivers' ability to recognize other hazards. (Interview with Dr. Michael Flannagan, 2008) Making an emergency vehicle "too conspicuous" could also lead to the driver fixation phenomenon described previously. Overdoing the use of retroreflective sheeting is probably not a major concern unless it is applied to the vehicle without attention to how it might be interpreted by approaching drivers. Additional research is needed to determine the effects of varying levels of conspicuity treatments under different scenarios (i.e., during an emergency response versus stopped along the roadside).

Battenburg Pattern

It appears the idea of alternating square blocks of contrasting fluorescent colors to increase vehicle visibility/conspicuity reaches back to a 1965 study titled "Development of a Paint Scheme for Increasing Aircraft Detectability and Visibility." (Siegel and Federman) Resembling the "harlequin" pattern recommended for law enforcement vehicles by U.S. researchers in a 1981 NIJ report, (Rubin & Howett) application of the Battenburg pattern to police vehicles across the United Kingdom was intended to enhance their visibility/ conspicuity and identification with law enforcement agencies. (Thomas, 1998; Harrison, 2004) Various configurations and colors of the Battenburg pattern were tested in experimental trials for the United Kingdom Police Scientific Development Branch (PSDB) from 1992 to 1998. (Harrison, 2004) These test results underpinned the ultimate specifications published by the PSDB and the United Kingdom Home Office Scientific Development Branch. (Thomas, 1998; Harrison, 2004, 2006)

Several studies examined for this report, however, expressed concern that the Battenburg pattern might actually hinder visibility by creating a camouflage effect, particularly against a visually-cluttered background. (De Lorenzo & Eilers, 1991; Langham & Rillie, 2002; Tijerina et al., 2003) Additional research is needed to determine if, on balance, the harlequin/Battenburg pattern aids or impedes emergency vehicle visibility and conspicuity. Without specific studies evaluating its real-world success at reducing crashes, the effectiveness of the Battenburg pattern in the United Kingdom appears primarily related to its widespread association with police vehicles in that country; this standardization effect likely alters drivers' actions in the United Kingdom upon recognizing an emergency vehicle. Since it remains a relative novelty in the United States, it is not clear that importing the full- or half-Battenburg pattern for use on U.S. emergency vehicles is a wise idea, especially given the present lack of research specific to its effects on visibility/conspicuity and driver behavior in the U.S. traffic system.

Opportunities

The consensus among studies and researchers cited in this report suggests there are a number of practical things that law enforcement agencies, EMS providers, and fire departments can do immediately to enhance the ability for other drivers to see and recognize emergency vehicles during all phases of an incident.

Contour Markings

Outlining vehicle boundaries with "contour" or "edge" markings, using retroreflective material, is expected to help enhance emergency vehicle visibility/conspicuity. The potential value of outlining a vehicle on its ultimate visibility/conspicuity is supported by research going back to 1984. (Henderson et al.) A Canadian study of large truck trailers identified continuous contour markings, made with white retroreflective tape, on the sides and rear of trailers to be more visible under varied weather conditions than the standard FMVSS 108 conspicuity treatment required by U.S. regulations. (Hildebrand & Fullarton, 1997) A 1999 United Kingdom study found that fully outlining (large) vehicles with contour markings increased surrounding drivers' ability to detect them both day and night, as well as judge their size and distance. (Cook et al.) In an extensive study of various marking schemes for large trucks, Darmstadt University of Technology researchers found contour markings useful for improving both side- and rear- visibility. (Schmidt-Clausen, 2000) Langham and Rillie (2002) explained the benefits of marking a vehicle to project its entire shape, a recommendation echoed by Tijerina, et al. (2003) for improving the safety of the Ford Crown Victoria

Police Interceptor (CVPI). Loughborough University researchers, in a United Kingdom Department for Transport analysis, evaluated the cost of line and contour markings of HGVs and buses, relative to their safety benefits. This study demonstrated measurable benefits for both schemes, while acknowledging the need for additional research. (Richardson & Loughton, 2005) **Figures 13 and 14** illustrate the contour and edge marking techniques, respectively.

Ready-to-install packages of retroreflective material for applying edge markings to common types of U.S. law enforcement and other emergency vehicles are now commercially available at a relatively low cost.

Placement

Vehicle lighting technology used in passenger and commercial vehicles is constantly evolving. As newer versions of vehicle headlamps are deployed, changes in the way they illuminate the road ahead, including traffic signs, people, and emergency vehicles, inevitably follow. Studies (Chrysler et al., 2003; Sivak et al., 2006) of recent changes in headlamp illumination suggest it might be efficacious to concentrate retroreflective material lower on emergency vehicles to optimize interaction with approaching vehicles' headlamps. This opportunity does not replace, but rather complements, the anticipated positive effects of contour markings outlining an emergency vehicle's overall size and shape. For law enforcement vehicles, retroreflective material can be concentrated on the sides and rear to maintain stealth when facing traffic or patrolling. Retroreflective tape matched to the vehicle's base color also can be used to maintain an unmarked appearance during the day, but enhance visibility/conspicuity at night.

Fluorescent Colors

Consider (and allow) the use of fluorescent retroreflective materials in applications where a high degree of day or nighttime visibility is desired. Zwahlen and Vel (1994) basically summarize the value of fluorescent colors for emergency (and other) vehicles:

> It is recommended that designers of traffic signs, personal conspicuity enhancement items and devices, and roadside traffic control devices consider the superior visual conspicuity properties of fluorescent colors (especially fluorescent yellow and fluorescent orange) and incorporate them in designs when the highest possible daytime target conspicuity is absolutely necessary. (abstract)

With respect to U.S. emergency vehicle schemes, the increasing use of fluorescent colors will likely prove beneficial for providing 24/7/365 high-conspicuity on fire apparatus and ambulances. Mission requirements for law enforcement vehicles should drive decisions about whether to incorporate fluorescent colors in their liveries. For example, a traffic enforcement vehicle designed to be inconspicuous will probably not use fluorescent colors to enhance its daytime visibility (although color-matched retroreflective edge markings should be considered for officer safety at night).

Efficiency

Using high-efficiency retroreflective material can improve conspicuity while reducing the amount of vehicle surface area requiring treatment. Recent studies of retroreflective sheeting types in traffic control applications (Carlson, 2001; Gates & Hawkins, 2004; Amjadi, 2008) suggest the cost increase to specify higher-efficiency retroreflective material can be reasonably expected to pay off by reducing crashes under some scenarios. That said, another study of traffic sign legibility concluded with this sound advice:

> When selecting sign material, all visual performance factors must be considered: detection, color recognition, shape recognition, and legibility. This performance should be evaluated for all lighting and weather conditions. In addition, durability, ease of fabrication, and cost must be weighed against the benefits of each product. (Chrysler et al., 2002, p.33)

Figure 13–Contour markings on large vehicles.

Figure 14–Edge markings on a patrol car.

This guidance is particularly important for emergency vehicle applications since installation, maintenance, and storage considerations can affect the long-run retroreflective performance of any given sheeting material.

Logos and Emblems

Applying distinctive logos or emblems made with retroreflective material could improve emergency vehicle visibility and recognition. European studies on the use of retroreflectorized logos and graphics found the application of simple designs made from retroreflective sheeting markedly improved the visibility/conspicuity of heavy trucks. (Schmidt-Clausen, 2000) The use of clearly identifiable logos or graphics specifying the affiliation, and therefore function, of an emergency vehicle can be reasonably expected to aid recognition and help surrounding drivers better anticipate its behavior; two examples made from high-efficiency microprismatic sheeting are displayed in **Figure 15**.

 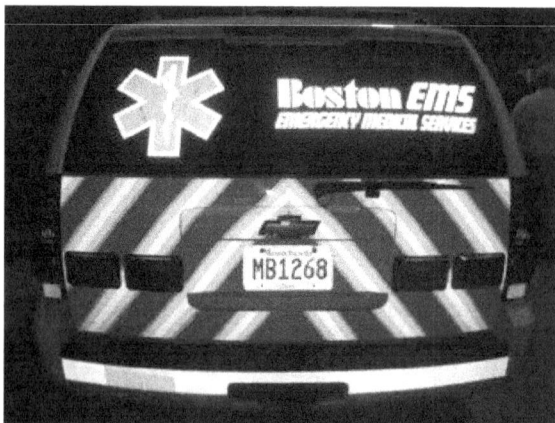

Figure 15–Retroreflective logos and emblems.

Conclusion

Advancing the state-of-the-art in emergency vehicle visibility/conspicuity will likely result from a combination of both active and passive conspicuity treatments—including enhanced emergency vehicle warning lighting systems and the increased use of retroreflective materials—to improve the visibility and recognizability (when desired) of emergency vehicles including ambulances, patrol cars, and fire apparatus.

Additional research specific to emergency vehicle visibility and conspicuity is critically needed in the United States, particularly since vehicle recognition is such a crucial facet of understanding how to improve responders' safety along the roadside.

Notwithstanding the importance of additional scientific study, it is noteworthy that many advances in vehicle and traffic safety over the years were successfully made using a "common sense" approach. (Sivak & Tsimhoni, 2008) For this reason, sensible efforts to improve the visibility and conspicuity of emergency vehicles need not be delayed; however, these efforts must be followed-up, in short order, with empirical studies to determine their effectiveness and identify any unintended consequences.

Photo Credits

Cover: Ambulance photo courtesy of Assistant Chief Jonathan Olson, Wake County Department of Emergency Medical Services, North Carolina; other photos by Adam Thiel

Figure 5–www.policecarsite.50webs.com

Figure 6–Wikimedia Commons

Figure 7–Wikimedia Commons and www.policecarsite.50webs.com, Ian Marlow

Figure 8–Wikimedia Commons

Figure 9–Lee Wilson

Figure 10–Arizona Department of Public Safety. Note: The reflectors provided as standard on this vehicle type are not visible in the lower photograph.

Figure 11–Montgomery County, Maryland, Fire and Rescue Service

Figure 12–Adam Thiel

Figure 13–Courtesy of Reflexlite Americas

Figure 14–Courtesy of Policelines.com

Figure 15–Julie Willett and Adam Thiel

References

3M. (2009). 3M Roadwise: Reflective materials-science-pupils. http://www.3m.com/intl/ie/3mroad wise/pupils-reflective-materials.htm. Accessed 2/15/09.

AAA. (2007). Improving Traffic Safety Culture in the United States: The Journey Forward. AAA Foundation for Traffic Safety: Washington, DC.

Amjadi, R. (2008). Techbrief: Safety evaluation of increasing retroreflectivity of stop signs. Federal Highway Administration: Washington, DC. Report # FHWA-HRT-08-047. March 2008.

Anders, R.L. (2000). On-road investigation of fluorescent sign colors to improve conspicuity. Virginia Polytechnic Institute and State University: Blacksburg, VA.

Aoki, T., Battle, D.S., and Olson, P.L. (1989). The subjective brightness of retroreflective sign colors: Final report. University of Michigan Transportation Research Institute: Ann Arbor, MI. UMTRI-89-22.

Arizona Department of Public Safety. (2003). Four Arizona Tragedies. International Association of Chiefs of Police: Alexandria, VA. http://www.theiacp.org/Portals/0/ppts/AZ_DPS/ AZ_DPS_files/frame.htm. Accessed 2/20/09.

ASTM International. (2007). Standard Specification for Retroreflective Sheeting for Traffic Control. ASTM International: West Conshohocken, PA. ASTM D4956-07e1.

_____. (2008). Standard Test Method for Coefficient of Retroreflection of Retroreflective Sheeting Utilizing the Coplanar Geometry. ASTM International: West Conshohocken, PA. ASTM E810-03 (2008).

Braver, E.R., Preusser, D.F., Williams, A.F., and Weinstein, H.B. (1996). Major types of fatal crashes between large trucks and cars. Insurance Institute for Highway Safety: Arlington, VA.

British Standards Institute. (2007). Medical vehicles and their equipment-road ambulances. BS EN 1789:2007.

Buonarosa, M.L., and Sayer, James R. (2007). Seasonal Variations in Conspicuity of High-Visibility Garments. University of Michigan Transportation Research Institute: Ann Arbor, MI. UMTRI-2007-42.

Burbank, A. (2007). Police officer visibility at roadside emergencies: A risk management perspective. New Hampshire Town and City. New Hampshire Local Government Center: Concord, NH. October 2007.

Burger, W.J., Mulholland, M.U., and Smith, R.L. (1985). Improved Commercial Vehicle Conspicuity and Signaling Systems. Vector Enterprises. National Highway Traffic Safety Administration: Washington, DC. Report# HS-806 923.

Burger, W.J., and Smith, R.L. (1987). Use of retroreflectorization to reduce truck-trailer accidents. Transportation Research Board: Washington, DC. Transportation Research Record, 1149.

Carlson, P.J. (2001). Evaluation of Clearview Alphabet with Microprismatic Retroreflective Sheeting. Texas Transportation Institute. Texas A&M University: College Station, TX. Report# 4049-1. October 2001.

Carlson, P.J., and Urbanik, T., II. (2004). Validation of photometric modeling techniques for retroreflective traffic signs. National Standards Council: Washington, DC. Transportation Research Record, 1862, 109-118.

Charles, M., Crank, J., and Falcone, D.N. (1990). A search for evidence of the fascination phenomenon in road side accidents. AAA Foundation for Traffic Safety: Washington, DC.

Chatziastros, A., Readinger, W., and Bülthoff, H. (2003). Environmental variables in the "moth effect." Vision in Vehicles X: Granada, Spain. September 2003.

Chrysler, S.T., Carlson, P.J., and Hawkins, H.G. (2002). Nighttime legibility of ground-mounted traffic signs as a function of font, color, and retroreflective sheeting type. Texas Transportation Institute. Texas A&M University: College Station, TX. Report# 0-1796-2.

_____. (2003). Headlamp illumination provided to sign positions by passenger vehicles. Texas Transportation Institute. Texas A&M University: College Station, TX. Report# 0-1796-3.

Cole, B.L., and Hughes, P.K. (1984). Field trial of attention and search conspicuity. Human Factors and Ergonomics Society. Human Factors, 26, 3.

Cook, S., Quigley, C., and Clift, L. (1999). An assessment of the contribution of retroreflective and fluorescent materials. U.K. Department of Environment, Transport, and the Regions.

Cumberland Valley Volunteer Firemen's Association. (1999). Protecting Emergency Responders on the Highways: A White Paper. United States Fire Administration: Emmitsburg, MD.

De Lorenzo, R., and Eilers, M. (1991). Lights & Siren: A review of emergency vehicle warning systems, Annals of Emergency Medicine, 20, 12, 1331-1335.

Donelson, A.C., and Ayres, T.J. (2007). Preventing Crashes with Trucks at Night: A Few Lessons from Experience. World Conference on Transport Research Society.

Dula, C.S., and Geller, E.S. (2007). Creating a Total Safety Traffic Culture. AAA Foundation for Traffic Safety: Washington, DC.

Eby, D. and Bingham, C.R. (2007). Customized Driver Feedback and Traffic-Safety Culture. AAA Foundation for Traffic Safety: Washington, DC.

European Committee for Standardization (CEN). (2007). Medical vehicles and their equipment-road ambulances. Brussels, Belgium. EN 1789:2007.

Federal Bureau of Investigation. (2007). Law Enforcement Officers Killed and Assaulted reports, 1996-2007. http://www.fbi.gov/ucr/ucr.htm#leoka. Accessed 3/24/09.

Federal Highway Administration. (2005). FHWA retroreflective sheeting identification guide-September 2005. http://safety.fhwa.dot.gov/roadway_dept/retro/sign/retrore_sheet_ id.htm. Accessed 2/15/09.

_____. (2007). Manual on Uniform Traffic Control Devices, 2003 Edition with Revisions Number 1 and 2 Incorporated. U.S. Department of Transportation: Washington, DC. December 2007.

_____. (2009). The physics of retroreflectivity. http://safety.fhwa.dot. gov/roadway_dept/retro/gen/back_physics.htm. Accessed 2/15/09.

Fontaine, M.D., and Hawkins, H.G. (2001). Catalog of effective treatments to improve driver and worker safety at short-term work zones. Texas Transportation Institute. Texas A&M University: College Station, TX. Report# 0-1879-3. January 2001.

Forbes, T.W. (1981). Practical aspects of conspicuity principles. Transportation Research Circular No. 229. Transportation Research Board: Washington, DC.

Foss, R. (2007). Addressing Behavioral Elements in Traffic Safety: A Recommended Approach. AAA Foundation for Traffic Safety: Washington, DC.

Flannagan, M.J., and Devonshire, J.M. (2007). Effects of warning lamps on pedestrian visibility and driver behavior. University of Michigan Transportation Research Institute: Ann Arbor, MI. April 2007.

Gates, T.J., and Hawkins, H.G. (2004). Effect of higher-conspicuity warning and regulatory signs on driver behavior. Texas Transportation Institute. Texas A&M University: College Station, TX. Report# 0-4271-S.

Girasek, D.C. (2007). Moving America Towards Evidence-Based Approaches to Traffic Safety. AAA Foundation for Traffic Safety: Washington, DC.

Green, P., Kubacki, M., Olson, P.L., and Sivak, M. (1979). Accidents and the nighttime conspicuity of trucks. University of Michigan Highway Safety Research Institute: Ann Arbor, MI. UM-HSRI-79-92.

Green, M. (2009). Visual expert human factors: Is the moth effect real? http://www.visualexpert.com/Resources/motheffect.html. Accessed 2/15/09.

Harrison, P. (2004). High-conspicuity livery for police vehicles. U.K. Home Office Police Scientific Development Branch: Hertfordshire, U.K. Publication No. 14/04.

_____. (2006). High-conspicuity livery for police motor cycles. U.K. Home Office Police Scientific Development Branch: Hertfordshire, U.K. Publication No. 47/06.

Harsha, B., and Hedlund, J. (2007). Changing America's Culture of Speed on the Roads. AAA Foundation for Traffic Safety: Washington, DC.

Hawkins, H.G., Carlson, P.J., and Elmquist, M. (2000). Evaluation of fluorescent orange signs. Texas Transportation Institute. Texas A&M University: College Station, TX. Report# 0-2962-S. May 2000.

Hildebrand, E.D., and Fullarton, P.J. (1997). Effectiveness of heavy truck conspicuity treatments under different weather conditions. University of New Brunswick, Canada.

Henderson, R.L., Ziedman, K., Burger, W.J., and Cavey, K.E. (1984). Motor vehicle conspicuity. SAE Technical Paper Series. Society of Automotive Engineers: Warrendale, PA.

Ho, G., Scialfa, C.T., Caird, J.K., and Graw, T. (2001). Visual search for traffic signs: The effects of clutter, luminance, and aging. Human Factors and Ergonomics Society. Human Factors, 43, 2, 194-207.

International Association of Chiefs of Police. (2004). Law Enforcement Stops and Safety Subcommittee (LESS): Staff Study 2004. International Association of Chiefs of Police: Alexandria, VA.

Killeen, J. (2008). Ambulance visibility issues. Transportation Research Board Ambulance Safety Summit. National Academies: Washington, DC. November 7, 2008.

Kitamura, F., and Matsunaga, K. (1994). Influence of looking at hazard lights on car-driving performance. Perceptual & Motor Skills, 78, 1059-1065.

Kleinschmit, M.W., and Couzin, D. (2007). Evaluation of road-sign geometry parameter space. Transportation Research Board. 18th Biennial Visibility Symposium: College Station, TX.

Knipling, R., Wang, J.S., and Yin, H.M. (1993). Rear-end crashes: Problem size assessment and statistical description. U.S. Department of Transportation, National Highway Traffic Safety Administration: Washington, DC. Report# DOT HS 807 994.

Krull, K.A., and Hummer, J.E. (2000). The effect of fluorescent yellow warning signs at hazardous locations. Southeastern Transportation Center. North Carolina State University: Raleigh, NC.

Kuemmel, D.A. (1992). Maximizing Legibility of Traffic Signs in Construction Work Zones. Transportation Research Board: Washington, DC. Transportation Research Record 1352, 25-34.

Langham, M., Hole, G., Edwards, J., and O'Neil, C. (2002). An analysis of "looked but failed to see" accidents involving parked police vehicles. Ergonomics, 45, 3, 167-185.

Langham, M., and Rillie, I. (2002). Making vehicles more conspicuous: The application of conspicuity theory. Transport Research Laboratory (TRL) Limited: Crowthorne, Berkshire, UK.

Lardelli-Claret, P., Luna-del-Castillo, J.D., Jimenez-Moleon, J.J., Femia-Marzo, P., Moreno-Abril, O., and Bueno-Cavanillas, A. (2002). Does Vehicle Color Influence the Risk of Being Passively Involved in a Collision? Epidemiology, 13, 6, 721-724.

Lloyd, J. (2009). A brief history of retroreflective sign face sheet materials. The Retroreflective Equipment Manufacturers Association: Lancashire, U.K. http://www.rema.org.uk/. Accessed 2/15/09.

Lum, H.S. (1979). Evaluation of techniques for warning off slow moving vehicles ahead: Executive summary. Federal Highway Administration: Washington, DC. Report# FHWA-RD-79-97.

Lyles, R.W. (1980). Effective warning devices for parked/disabled vehicles: Executive summary. Federal Highway Administration: Washington, DC. Report# FHWA-RD-80-064.

Maguire, B.J., Hunting, K.L., Smith, G.S., and Levick, N.R. (2002). Occupational fatalities in emergency medical services: A hidden crisis. Annals of Emergency Medicine, 40, 6, 625-632.

McNeely, C.L., and Gifford, J.L. (2007). Effecting a Traffic Safety Culture: Lessons From Cultural Change Initiatives. AAA Foundation for Traffic Safety: Washington, DC.

McCann, H., Averitt, R., and Williams, E. (2008). Lights and stripes: Proper use could save your life. EHS Today. Penton Media: New York, NY. October 2008.

Minimumreflectivity.org. (2009). Guidance for improving roadway safety, understanding minimum retroreflectivity standards: Science of retroreflectivity. http://www.minimum reflectivity.org/retroreflective.asp. Accessed 2/15/09.

Morgan, C. (2001). The effectiveness of retroreflective tape on heavy trailers. National Highway Traffic Safety Administration: Washington, DC.

Mortimer, R. (1990). Perceptual factors in rear-end crashes. Proceedings of the Human Factors Society 34th Annual Meeting. 591-594.

National Fire Protection Association. (2009). NFPA 1901—Standard for automotive fire apparatus, 2009 edition. Quincy, MA.

National Law Enforcement Officers Memorial Fund. (2008). Law enforcement officer deaths: Preliminary 2008 report. December 2008 Research Bulletin. Washington, DC.

Olson, P., Campbell, K., Massie, D., Battle, D.S., Traube, E.C., Aoki, T., Sato, T., and Pettis, L.C. (1992). Performance requirements for large truck conspicuity enhancements. University of Michigan Transportation Research Institute: Ann Arbor, MI. UMTRI-92-8.

Olsen, R.A. (1981). Providing for visibility in night driving. Transportation Research Circular 229. Transportation Research Board: Washington, DC.

Organization for Economic Cooperation and Development. (1988). The role of heavy freight vehicles in traffic accidents: Report on the symposium held in Montreal, April 1987.

Owens, D.A., and Sivak, M. (1993). The role of reduced visibility in nighttime road fatalities. University of Michigan Transportation Research Institute: Ann Arbor, MI. UMTRI-92-8. UMTRI-93-33.

Readinger, W.O., Chatziastros, A., Cunningham, D.W., Bülthoff, H.H., and Cutting, J.E. (2002). Gaze-eccentricity effects on road position and steering. Journal of Experimental Psychology: Applied. 8, 4, 247-258.

Richardson, J., and Lawton, C. (2005). The safety benefit of retroreflective markings on HGVs and buses: Partial RIA - preliminary report. UK Department for Transport. Loughborough University: Loughborough, UK. June 2005. http://hdl.handle.net/2134/552

Ridenour, M., Noe, R.S., Proudfoot, S.L., Jackson, J.S., Hales, T.R., and Baldwin, T.N. (2008). NIOSH Fire Fighter Fatality Investigation and Prevention Program: Leading recommendations for preventing fire fighter fatalities, 1998-2005. National Institute for Occupational Safety and Health: Washington, DC.

Rogoff, M.J., Rodriguez, A.S., and McCarthy, M.B. (2005). Using retroreflectivity measurements to assist in the development of a local traffic sign management program. Institute of Traffic Engineers: Washington, DC. ITE Journal, 75, 10, 28-32.

Rubin, A., and Howett, G.L. (1981). Emergency Vehicle Warning Systems. National Institute of Justice: Rockville, MD.

Saunders, G., and Gough, A. (2003). Emergency ambulances on the public highway linked with inconvenience and potential danger to road users. Emergency Medical Journal, 20, 277-280.

Sawyer, D., Stewart, R., and Talley, W.T. (1988). Familiarity effects of daytime rear running lights. AAA Foundation for Traffic Safety: Washington, DC.

Sayer, J.R., and Mefford, M.L. (2007). The Roles of Garment Design and Scene Complexity in the Daytime Conspicuity of High-Visibility Safety Apparel. National Association of Professional Accident Reconstruction Specialists. Accident Reconstruction Journal, 17, 5, 51-54.

Schieber, F., Willan, N., and Schlorholtz, B. (2003). Fluorescent colored stimuli automatically attract visual attention: An eye movement study. University of South Dakota: Vermillion, SD.

Schreuder, D.A. (1985). Visibility aspects of road lighting. Transportation Research Circular 297. Transportation Research Board: Washington, DC.

Schmidt-Clausen, J. (2000). Retroreflective marking of vehicles. Darmstadt University of Technology: Munich, Germany.

Siegel, A., and Federman, P. (1965). Development of a Paint Scheme for Increasing Aircraft Detectability and Visibility. Journal of Applied Psychology, 49, 2, 93-105.

Sivak, M. (1979). A review of literature on nighttime conspicuity and effects of retroreflectorization. Highway Safety Research Institute: Washington, DC. HSRI Research Review, 10, 3.

_____. M. (1987). Human factors and road safety: Overview of research at the University of Michigan Transportation Research Institute between 1977 and 1986. University of Michigan Transportation Research Institute. UMTRI-87-1.

Sivak, M., Soler, J., Trankle, U., and Spagnhol, J.M. (1989). Cross-cultural differences in driver risk-perception. Accident Analysis and Prevention 21, 4, 355-362.

Sivak, M., Schoettle, B., and Flannagan, M.J. (2006). Recent changes in headlamp illumination directed toward traffic signs. University of Michigan Transportation Research Institute: Ann Arbor, MI. UMTRI-2006-31.

Sivak, M., and Tsimhoni, O. (2008). Improving traffic safety: Conceptual considerations for successful action. University of Michigan Transportation Research Institute: Ann Arbor, MI. UMTRI-2008-21.

Smith, H.J. (1981). Daylight fluorescent color—the color that shouts. Transportation Research Circular 229. Transportation Research Board: Washington, DC.

Smith, K., and Martin, J.W. (2007). A Barrier To Building A Traffic Safety Culture In America: Understanding Why Drivers Feel Invulnerable And Ambivalent When It Comes To Traffic Safety. AAA Foundation for Traffic Safety: Washington, DC.

Solomon, S.S. (1990). Lime-yellow color as related to reduction of serious fire apparatus accidents: The case for visibility in emergency vehicle accident avoidance. Journal of the American Optometric Association, 61, 827-831.

Solomon, S.S., and King, J.G. (1995). Influence of color on fire vehicle accidents. Journal of Safety Research, 26, 1, 41-48.

_____. (1997). Fire truck visibility. Human Factors and Ergonomics Society: Santa Monica, CA. Ergonomics in Design, 5, 2.

Stutts, J.C., Reinfurt, D.W., Staplin, L., and Rodgman, E.A. (2001). The role of driver distraction in traffic crashes. AAA Foundation for Traffic Safety: Washington, DC.

Sullivan, J.M. (2005). Further crash evidence on the nighttime visibility of trucks. University of Michigan Transportation Research Institute: Ann Arbor, MI. UMTRI-2005-22.

Sullivan, J.M., and Flannagan, M.J. (2004). Visibility and rear-end collisions involving light vehicles and trucks. University of Michigan Transportation Research Institute: Ann Arbor, MI. UMTRI-2004-14.

Sweatman, P.F. (1991). Review of vehicle factors in truck crashes: Australian truck safety study task 2. Australian Road Research Board: Vermont South, Victoria. Research Report# ARR 202.

Texas Transportation Institute. (2004). How retro is your reflectivity: TTI research contributes to national standards. Texas A&M University: College Station, TX. Texas Transportation Researcher, 40, 1, 4-7.

Thomas, A. (1998). Specification for the livery on police patrol cars. U.K. Home Office Police Policy Directorate, Police Scientific Development Branch: Hertfordshire, U.K. Publication No. 2/98.

Tijerina, L., Shulman, M., Wells, J.D., and Kochhar, D. (2003). Committee report: Conspicuity enhancement for police interceptor rear-end crash mitigation. https://www.fleet.ford. com/showroom/CVPI/pdfs/ CVPI_Conspicuity_Report.pdf. Accessed 2/15/09.

Tutterow, R. (2008). Chevrons on the rear of fire apparatus: Background. NFPA Fire Service Section Newsletter. National Fire Protection Association: Quincy, MA. Winter 2008.

TUV Rheinland Group. (2004). Conspicuity of heavy goods vehicles. European Commission, Directorate General for Energy and Transport.

Ullman, G.L., Ullman, B.R., and Finley, M.D. (2005). Evaluating the safety risk of active night work zones. Texas Transportation Institute. Texas A&M University: College Station, TX. Report# 0-4747-2. April 2005.

United Kingdom-Parliament. (1989). The Road Vehicles Lighting Regulations 1989. Statutory Instrument 1989 No. 1796. Crown Copyright: London, U.K.

U.K. Department of Transport. (2008). Impact assessment of the Road Vehicle Lighting Regulations amendment covering reflective markings on emergency vehicles. March 7, 2008.

U.S. General Services Administration. (2007). Federal specification for the star-of-life ambulance. KKK-A-1822F-08. U.S. General Services Administration: Washington, DC.

U.S. Code of Federal Regulations. (2004). Title 49, Part 571, Standard 108 Federal Motor Vehicle Safety Standards. National Highway Traffic Safety Administration: Washington, DC.

United States Fire Administration. (2002). Firefighter Fatality Retrospective Study. April 2002: Emmitsburg, MD. Report FA-220.

_____. (2004). Emergency Vehicle Safety Initiative. August 2004: Emmitsburg, MD. Report FA-272.

_____. (2005). Study of Emergency Vehicle Warning Lighting: Inferences about Emergency Vehicle Warning Lighting Systems from Crash Data. July 2005.

_____. (2008). Firefighter Fatalities in the United States in 2007. June 2008: Emmitsburg, MD.

_____. (2009a). USFA Releases Provisional 2008 Firefighter Fatality Statistics. January 7, 2009. Emmitsburg, MD. http://www.usfa.dhs.gov/media/press/2009releases/ 010709.shtm. Accessed 2/15/2009.

_____. (2009b). Emergency Vehicle Safety. Emmitsburg, MD. http://www.usfa.dhs.gov/ fireservice/research/safety/vehicle.shtm. Accessed 2/15/09.

_____. (2009c). Chart illustrating seeing and stopping distances. Emmitsburg, MD. http://www.usfa.dhs.gov/fireservice/research/safety/vehicle.shtm. Accessed 2/15/09.

Ward, N.J. (2007). The Culture of Traffic Safety in Rural America. AAA Foundation for Traffic Safety: Washington, DC.

Wickens, C.D., and Hollands, J. (2000). Engineering psychology and human performance, 3rd edition. Prentice-Hall: New York, NY.

Wiegmann, D.A., Thaden, T.L., and Gibbons, A.M. (2007). A Review of Safety Culture Theory and Its Potential Application to Traffic Safety. AAA Foundation for Traffic Safety: Washington, DC.

Williams, A.F., and Haworth, N. (2007). Overcoming Barriers to Creating a Well-Functioning Safety Culture: A Comparison of Australia and the United States. AAA Foundation for Traffic Safety: Washington, DC.

Zwahlen, H.T., and Vel, U.D. (1994). Conspicuity in terms of peripheral visual detection and recognition of fluorescent color targets versus nonfluorescent color targets against different backgrounds in daytime. Transportation Research Record 1465. Transportation Research Board: Washington, DC.

Appendix A. FHWA Retroreflective Sheeting Identification Guide (FHWA, 2005)

FHWA Retroreflective Sheeting Identification Guide – September 2005

Notes: ASTM Types are shown as stated by the manufacturers using ASTM D4956-04 "type" designations.
Agencies should verify that the sheeting they use complies with their specifications or ASTM D4956.
FHWA does not endorse or approve any material nor does it determine type category(s) for materials.
This side of the Sheeting ID Guide is for rigid surfaces only. The other side is for flexible surfaces and non-signing applications.

Retroreflective Sheeting Materials for Rigid Sign Surfaces Made with Glass Beads

Example of Sheeting (Shown to scale)									
ASTM Type	I	II	II	III	III	III	III	III	III
Manufacturer	See note A	Avery Dennison®	Nippon Carbide	3M™	ATSM, Inc.	Avery Dennison®	Kiwalite®	LG Lite	Nippon Carbide
Brand Name	Engineer Grade	Super Engineer Grade	Super Engineer Grade	High Intensity	High Intensity	High Intensity	High Intensity	High Intensity	High Intensity
Series Number	Several	T-2000	15000 17000 18000	2800 3800	ASTM HI	T-5500	22000	LH8000 LH8100	N500 N800
NOTES:	A								

Retroreflective Sheeting Materials for Rigid Sign Surfaces Made with Prisms

Example of Sheeting (Shown to scale)									
ASTM Type	III, IV	III, IV, X	VII, VIII, X	VIII	IV, VIII	IX	IX	X	Unassigned
Manufacturer	Avery Dennison®	3M™	3M™	Avery Dennison®	Nippon Carbide	3M™	Avery Dennison®	Nippon Carbide	3M™
Brand Name	High Intensity Prismatic	High Intensity Prismatic	Diamond Grade™ LDP	MVP Prismatic	Crystal Grade	Diamond Grade™ VIP	Omni-View™	Crystal Grade	Diamond Grade™ DG3
Series Number	T-6500	3930	3970	T-7500	94000 (IV) 92000 (VIII)	3990	T-9500	93000	4000
NOTES:	B	B	B,D	B,C				C	

A – All the manufacturers listed on the other side of this guide (except Reflexite) provide Engineer Grade sheeting. Engineer Grade sheeting is uniform without any patterns or identifying marks. Visually, it is indistinguishable from lower quality grades (i.e., utility and commercial grades).
B – These materials can be classified as different ASTM Types.
C – These materials are visually indistinguishable from one another.
D – The arrow or "water mark" on this product is no longer included with new productions.

FHWA Retroreflective Sheeting Identification Guide – September 2005

Notes: ASTM Types are shown as stated by the manufacturers using ASTM D4956-04 "type" designations.
Agencies should verify that the sheeting they use complies with their specifications or ASTM D4956.
FHWA does not endorse or approve any material nor does it determine type category(s) for materials.
This side of the Sheeting ID Guide is for flexible and non-signing applications. The other side is for rigid surfaces.
Below are symbols that have been used to indicate special applications for sheeting on this side of the Sheeting ID Guide:

Cone Drum Temporary Tubes Sign

Retroreflective Sheeting Materials for Non-Signing Applications

Example of Sheeting (Shown to scale)						
ASTM Type	III	III	V	V	III	VI
Manufacturer	Avery Dennison®	Reflexite	Reflexite	Reflexite	3M™	Reflexite
Brand Name	High Intensity Prismatic Work Zone	High Impact Channelizer Tape	Barrier Delineator	Barrier Delineator	High Intensity Flexible	Traffic Cone Collar
Series Number	WR-6100	n/a	AR1000	AP1000	3840	n/a
Typical Use	Reboundable Device	Reboundable Device	Rigid Non-Signing Surface	Rigid Non-Signing Surface	Reboundable Device	Traffic Cone

Retroreflective Sheeting Materials for Flexible Signs

Example of Sheeting (Shown to scale)						
ASTM Type	VI	VI	VI	VI	VI	VI
Manufacturer	3M™	3M™	Avery Dennison®	Reflexite	Reflexite	Reflexite
Brand Name	Diamond Grade™ Roll-Up Sign	Vinyl Roll-Up Sign	Flexible Roll-Up Sign	Flagging Material	High Performance Marathon	Super Bright Fluorescent
Series Number	RS20	RS30	WU-6014	n/a	n/a	n/a
Typical Use	Roll-Up Sign	Roll-Up Sign	Roll-Up Sign	Roll-Up Sign	Roll-Up Sign	Roll-Up Sign

Contact Information

3M - www.3M.com/tcm	Kiwalite - www.kiwa-lite.com	Reflexite - www.reflexite.com
ATSM, Inc. - www.atsminc.com	LG Lite - www.lgchem.com	Nippon Carbide - www.nikkalite.com
Avery Dennison - www.reflectives.averydennison.com		FHWA - www.fhwa.dot.gov/retro

Source: http://safety.fhwa.dot.gov/roadway_dept/retro/sign/retrore_sheet_id.htm

Appendix B. Chevrons on the Rear of Fire Apparatus: The Background (Tutterow, 2008)

Chevrons on the Rear of Fire Apparatus: The Background
By Health & Safety Officer Robert Tutterow

The following information is a bit of the background on the chevron striping soon coming on all new fire apparatus.

The NFPA Technical Committee on Apparatus has developed requirements in the next revision of NFPA 1901—Standard for Automotive Fire Apparatus to address "conspicuity." However, the new requirements are for "on scene" safety rather than "responding" safety. The effective date is for all fire apparatus contracted on or after January 1, 2009.

One of the most controversial new requirements is chevron striping on the rear of apparatus. The idea of mandating the striping met virtually no opposition. However, there was considerable opposition to specifying the exact colors and size of the striping. I will explain the substantiation about that later. First, what exactly are the requirements of the chevron striping?

At least 50 percent of the rear-facing vertical surfaces, visible from the rear of the apparatus, excluding any pump panel areas not covered by a door, shall be equipped with retroreflective striping in a chevron pattern sloping downward and away from the center line of the vehicle at an angle of 45 degrees.

Each stripe in the chevron shall be a single color alternating between red and either yellow, fluorescent yellow, or fluorescent yellow green. Each stripe shall be 6 inches in width.

We cannot forget that conspicuity is not just a night-time issue. That is the primary reason that yellow was selected as one of the two colors. Yellow is the most distinguishable color in daylight, especially when framed by a sea of a concrete multi-lane highway. The IAFF has a vehicle safety program PowerPoint presentation. One of the slides includes a photo of a Plano Texas Fire Department on the scene of a MVA on a multi-lane road during sunlight. (The Plano Fire Department has used red/yellow chevron striping for years.) The Plano apparatus visually "pops out" among everything else in the photo. (See photo accompanying this article.)

As stated earlier, there was considerable opposition to specifying the colors and sizes. Many departments expressed their desire to choose their own colors and sizes—just as they have a choice in the color of their apparatus. The committee weighed this position very carefully and finally reached a consensus that a "standard is not a standard unless it is standard." As more and more of our incidents are highway incidents, the committee decided we should have a standard "look" when operating on roadways.

Fire departments must realize that roadway incidents involve a lot more than non-roadway incidents. Roadway incidents are typically multi-agency response incidents and traffic control is crucial. Far too many responders are being killed and injured while operating at roadway incidents and far too many apparatus are being struck while positioned at roadway incidents.

The motoring public needs to have consistent warnings across the Nation. Think about taking a family vacation to the beach or mountains. Your road trip is 4-6 hours. How many different fire response districts will you cross during this trip? What are the chances that more than one fire department will respond to a roadway emergency incident along the way? As a motorist, you expect the signage (and warnings) along the way to be consistent. What would you do if you came to a stop sign that was triangular in shape with white letters on a blue background? Stop signs are the same shape and color in every State. Yet, as a kid, I remember seeing yellow stop signs with black lettering. A consistent national average has its safety benefits. There is a reason the MUTCD included a *Chapter 61, Control of Traffic Through Traffic Incident Management Areas*. And, there is an organization known as NTIMC—*National Traffic Incident Management Coalition* which promotes a national agenda for standardized traffic incident management. Counter to the opposition, the committee had strong letters of support from the Health and Safety Section of the IAFC and the Emergency Responder Safety Institute strongly supporting the new requirement. It is for these reasons, and others, that the technical committee made the tough (and correct decision) to establish a standard. If a standard is not established now, it will be harder to establish one later and there might be an imitative for a non-fire service organization to develop our requirements—something we definitely do not need.

Reproduced with permission from the NFPA Fire Service Section Newsletter, Winter 2008, Copyright© 2008, National Fire Protection Association. This reprinted material is not the complete and official position of the NFPA on the referenced subject, which is represented only by the standard in its entirety.

Appendix C. NFPA 1901, Section 15.9.3 et seq. (NFPA, 2009)

15.9.3* Reflective Striping.

15.9.3.1* A retroreflective stripe(s) shall be affixed to at least 50 percent of the cab and body length on each side, excluding the pump panel areas, and at least 25 percent of the width of the front of the apparatus.

15.9.3.1.1 The stripe or combination of stripes shall be a minimum of 4 in. (100 mm) in total width.

15.9.3.1.2 The 4 in. (100 mm) wide stripe or combination of stripes shall be permitted to be interrupted by objects (i.e., receptacles, cracks between slats in roll up doors) provided the full stripe is seen as conspicuous when approaching the apparatus.

15.9.3.1.3 A graphic design shall be permitted to replace all or part of the required striping material if the design or combination thereof covers at least the same perimeter length(s) required by 15.9.3.1.

15.9.3.2 At least 50 percent of the rear-facing vertical surfaces, visible from the rear of the apparatus, excluding any pump panel areas not covered by a door, shall be equipped with retroreflective striping in a chevron pattern sloping downward and away from the centerline of the vehicle at an angle of 45 degrees.

15.9.3.2.1 Each stripe in the chevron shall be a single color alternating between red and either yellow, fluorescent yellow, or fluorescent yellow-green.

15.9.3.2.2 Each stripe shall be 6 in. (150 mm) in width.

15.9.3.3 All retroreflective materials required by 15.9.3.1 and 15.9.3.2 shall conform to the requirements of ASTM D 4956, *Standard Specification for Retroreflective Sheeting for Traffic Control,* Section 6.1.1 for Type I Sheeting.

15.9.3.3.1 All retroreflective materials used to satisfy the requirements of 15.9.3.1 that are colors not listed in ASTM D 4956, Section 6.1.1, shall have a minimum coefficient of retroreflection of 10 with observation angle of 0.2 degrees and entrance angle of -4 degrees.

15.9.3.3.2 Fluorescent yellow and fluorescent yellow-green retroreflective materials used to meet the requirements of 15.9.3.2 shall conform to the minimum requirements specified for yellow Type I Sheeting in ASTM D 4956, Section 6.1.1.

15.9.3.3.3 Any printed or processed retroreflective film construction used to meet the requirements of 15.9.3.1 and 15.9.3.2 shall conform to the standards required of an integral colored film as specified in ASTM D 4956, Section 6.1.1.